INDIAN ASTROLOGY LESSONS
(Self – Study Series – 2)

ASTRONOMY
relevant to
ASTROLOGY

Mullappilly Parameswaran

CreateSpace - Book ID : 5404483

Category   ;   Astrology

Name of the book   :   ASTRONOMY RELEVANT TO ASTROLOGY
(Indian astrology, self-study series 2)

Language of the book   : English

Author   :   Mullappilly Parameswaran
nullappillymp@gmail.com

First Edition   : 2015

ISBN (CreateSpace) :

        13: 978 - 1511508124
        10:     1511508124

Cover design and printing : CreateSpace
Published by the author
availing the self-publishing programme of CreateSpace

ASTRONOMY   RELEVANT   TO   ASTROLOGY

SYLLABUS
1.   What is astronomy?
2.   Earth and sky
3.   Planetary system
4.   Inner and outer planets
5.   Rahu and Ketu
6.   Comets
7.   Other Upagrahas
8.   The Sun and the stars
9.   Ecliptic
10.   Equator
11.   Longitude and latitude
12.   Sidereal and tropical zodiacs
13   Zodiac and constellations
14.   Planets considered in Hindu astronomy for astrplogical purposes
15.   The Luni-solar year
16.   The meani ng of Panchanga
17.   Eclipses.

This book is dedicated to:

my parents Mullappilly Krishan Elayalath, Kuzhur

and

Cherulli Nangeli Antherjanam, Chittanda, Trichur, Kerala, India)

LESSON - 1

ASTRONOMY *(JYOTISSASTRA)*

Definition:

Astronomy is the science of heavenly bodies such as the Sun, the Moon, planets, stars etc. It includes:

 a) computing the longitudes of planets, stars etc.,

 b) determining the time of eclipses,

 c) measurements of various celestial bodies; and

 d) their internal and external peculiarities etc.

INDIAN ASTRONOMICAL EPHEMERIS

Meteorological Department is publishig the Indian Astrono-mical Ephemeris every year. It is a store-house of data and contains all informations required for an Astronomer. Part one contains: calendar, ephemeris of the Sun, the Moon and all the planets including Pluto. Part two is about stars. There are six parts--all very useful. It also contains an Indian Calendar.

WESTERN ASTRONOMY

Out of the list of Western Astronomers, given in the Text Book, four names are to be recalled for their eminence and giving new revolutionary laws, discarding the age-old, ancient theories. They are Copernicus, Galileo, Kepler and Newton. In those days when religion was supreme, orthodoxy was rampant, and traditions were accepted without question, the men who founded the new astronomy must have had immense patience in observing and recording the observed values, besides great courage in framing them into hypotheses and laws, diametrically opposite to the views, then prevailed.

1. Copernicus (1473 - 1543) said in his theory, that the sun is the centre around which the earth and other planets revolve.

2. Galileo (1564-1642) held the same view, but he observed the planets and the moon with the aid of newly invented and improved telescopes, and gave some ideas of their nature.

3. Kepler (1571 - 1630), after studying carefully the records (made by his master for 20 years on the different positions of the planets of their movement in the firmament) gave three laws, governing their periods of revolution and the elliptical path they followed around the sun.

KEPLER'S LAWS

1.   State Kepler's Law of the Planetary motion. 12/2002.
2.   Define Kepler's Laws. 12/2000.

The Laws according to which the planets move around the Sun were discovered by John (Johannes) Kepler (1571 - 1630), a German scientist, as given below:

1.   Each planet moves in an elliptic orbit with the sun in one of the foci.

2.   Equal areas are covered in equal times by the radius of the planet, ie., by the line joining the planet and the sun.

3.   The squares of periodic times of the planets are to be one another as the cubes of their distances from the sun.

4.   Newton (1642 - 1772), confirmed these conclusions beyond doubt. He further summed up that all the laws of Kepler can be put into a single universal law for all bodies in the universe that every body attracts each other body with a force proportional

(1) directly to the product of t heir masses and

(2) inversely to the square of the distance between them.

POINTS OF DIFFERENCES BETWEEN INDIAN AND WESTERN ASTRONOMIES

1.   Distinguish between Indian and western astronomy. 6/98

2.   What are the differences between approaches of Indian astronomy and western astronomy? 12/2000

3.   Does the classical Indian astonomy differ with the modern western astronomy. Discuss. 12/2001

4.   What are the differences between modern western astronomy and Indian classical astronomy?   Discuss. 6/2003

1.   The western astronomical calculations are heliocentric, ie., based on the Sun. (The sun stationary and all the planets moving round it in somewhat elliptic orbits.).The Indian classical system, on the other hand, is Geocentric with the observer on the earth as centre.

2.   In Indian system, the time is reckoned from the beginning of the Universe and a number of revolutions considered for the planets in a Mahayuga of 4,320,000 solar years.   No such absolute motion from the time of the commencement of the Universe is followed in the western astronomy.

3.   The zodiac in western astronomy is Tropical or Moving where as the zodiac as per Indian system is Sidereal or Fixed. In *"Chitra paksha ayanamsa"* system of Indian astronomy, the zero point is a fixed point180o exactly opposite the star Chitra, ie., in between the two stars Revati and Aswini.   The western astronomers follow a moving "first point of Aries" which is now in Pisces 61/4o, that is, western calculation is 233/4o ahead of us.

Example:

Longitude of the Sun on 26-3-2004, 5-30 AM:
(Chitra paksha)   -   341o -   46'
Ayanamsa   (Chitrapaksha) -   + 23   -   55
Total   -   364 - 101   = 365-41
Western   (365.41 – 360)   5° - 41'

In other words, the western astronomy follows the *sayana* longitudes of the heavenly bodies from a point which is always moving westward @ 501/3 " per year,while Indian astrology follows sidereal longitudes from a fixed point , ie., a particular star.   It is why the Vasantha Vishu is on 20/21 March and the Vishu in Kerala is on April 14/15.

4.   The system of measurement of time is also different:

a) Day is related to the Earth's spin on its axis and in Indian system it is measured from <u>sunrise to next sunrise</u>. The western day is from midnight. to midnight.

 b) <u>Lunar month</u>.   Indian month is related to the Moon's motion round the Earth and its phases.   It is measured from Amavasi to next Amavasi,   ie., Sun and Moon same longitude.

(Kerala follows solar months).

 c) <u>Solar year.</u> We have civil days, Lunar months and Solar years of about 3651/4 days with months divided into 30/31/28/29 days and each day divided into 24 hours. Year is measured with Sun's apparent motion from Equinox to Equinox.

 The western day is from midnight to midnight and the days are fixed for each month. (January, 31 etc.)

5.   Indian system followed observations by the naked eye and calculations were made by using structures like "Sun-Dial" etc.

The western astronomy followed Kepler's laws of planetary motion, modern mathematics and detailed calculations.   Telescopes of very high resolution, photographic records from satellites, use of spectroscopes etc. have taken modern western astronomy far ahead of our classical system.

6.   In India, astronomy and astrology developed as a twin science.   In fact, the astronomical studies were accelerated for astrological purposes. But in the west, astronomy develped as a physical science

7.   In Indian system, some calculation points like Rahu and Ketu, Mandi, Yamakantaka, Karanas, Nityayogas etc. are considered which have no place in modern western astronomy.

CONTRIBUTIONS OF
CLASSICAL INDIAN ASTRONOMERS

Names of some of the famous Indian astronomers and mathematicians with their major works are given below. The dates in brackets refer either to the years of their birth (marked b.) or to the approximate years of composition of their major works.

S. No.	Author	Works
1.	Arybhata - I (b. 476 A.D.)	Aryabhatiyam, Aryasiddhanta
2.	Varahamihira (505 A.D.)	Panchasiddhantika, Brihatsamhita
3.	Bhaskara - I (c. 600 A.D.)	Bhasya on Aryabhatiyam, Mahabhaskariyam, Laghubhaskariyam
4.	Brahmagupta (b. 591 A.D.)	Brahmasphuta siddhanta, Khandakhadyaka
5.	Vateswara (880 A.D.)	Vateswarasiddhanta
6.	Manjula (932 A.D.)	Laghumanasam
7.	Aryabhata - II (950 A.D.)	Mahasiddhanta
8.	Bhaskara - II (b. 1114 A.D.)	Siddhanta siromani, Karanakutuhalam
9.	Parameswara (c. 1400 A.D.)	Drigganitam, Suryasiddhanta vivaranam, Bhatadipika etc.
10.	Nilakanta Somayaji (1465 A.D.)	Tantrasangraha, Aryabhatiya bhasya
11.	Ganesa Daivajna (1520 A.D.)	Grahalaghavam, Tithi Chintamani, Buddhi Vilasini

12.	Jyeshtadeva (1540 A.D.)	Yuktibhasa
13.	Chandrasekhara Samata (b.1835 A.D)	Siddhanta darpanah
14.	Sankara Varman (19th cent.)	Sadratnamala
15.	Venkatesa Ketkar (b. 1853 A.D.)	Grahaganitam, Jyotirganitam
16.	Venkata Ramana (1884) -	Ayanamsa Tattva Viveka

(Courtsey: Rashtriya Sanskrit Vidyapeetha, Tirupati)

LESSON - 2

AYANAMSA

As per the syllabus, the topic "Ayanamsa" comes within the next part only but it is discussed here to clear the confusions regarding some technical terms in respect of the Zodiac (*Rasi chakra*) and longitude of planets (*Graha-sphuta*) like Tropical – Sidereal / Sayana – Nirayana / Movable - Fixed etc.

1.   Ayanamsha is considered as a vexed problem of Hindu Astrology.   What is "Ayanamsha" and what are the kinds of problems one has to cope with on account of differences to its value.

2.   What is Ayanamsa? What, in your opinion, is the rate per year. If 285 A.D. is the year, when the Zero point was noted, what is ayanamsa for January1, 1991?   If the year is 397, what is the value   on January 1, 1991?

ZODIAC *(Rasi chakra, Rasi mandala)*

The Zodiac is an imaginary belt in the heavens, stretching about 8o - 9o either side, ie., north and south, of the ecliptic, within which the planets move around the Earth, once in a day, from east to west. (The movement of the planets and the zodiac from east to west is apparent; it so appears due to the rotation of the Earth from west to east.)

FIRST POINT OF ARIES
(*Meshadi*)

Since the zodiac is a circle with no beginning or end, an arbitrary point has been chosen to measure the 'signs' and also the longitudes of the planets. This point is known as the 'First Point of Aries' or Aries zero degree. As this is the point where the Sun enters northern hemisphere, it is also known as 'Vernal Equinox'.

EQUINOXES
(Ayanam, vishuvam, sama ratri dina)

Twice a year, ie., during his journey from south to north (*Uttar-ayana*) and from north to south (*Dakshinayana*), the Sun crosses the eqator (*Bhoomadhya rekha*). On these two days, the duration of day and night will be equal *(sama rata dina)*. These are known as Vernal and Autumnal equinoxes.

VERNAL EQUINOX

The Vernal Equinox (*Vasantha Vishu* or *Mahavishuvam*) is the *sama-ratri dina* during the Sun's movement from South to North (*Uttarayana*. from January, 14 to June, 15 or Malayala solar month *Makara* 1 to *Mithuna* 31. When the apparent longiude of the Sayana Sun is 0o, it is Vernal Equinox (about March 20). In the solar month of Kerala it is *Mesha Vishu Samkramam* (Medam 1) on April 14. The difference of 24 days (March, 20 to April, 14) is due to the *Chitra Paksha Ayanamsha.*

AUTUMNAL EQUINOX
During the journey from north to south, when the longitude of the Sun is 180o, it is Autumnal Equinox *(Thula vishu).* (about September 22).

Eg: Equinoxes and Solstices in 2004
(Lahiri's Ephemeris 2004. Page No.10)
Longitude of the Sun

	Sayana		Nirayana
Vernal Equinox	0o	20- 3-2004	14- 4-200
Summer Solstice	90o	21- 6-2004	16- 7-200
Autumnal Equinox	180o	22- 9-2004	16-10-2004Winter
Solstice	270o	21-12-2004	14- 1-2005

PRECESSION OF THE EQUINOXES *(Ayana chalanam)*

At the time when the Sun reaches the equinoctial point at Aries 0o , the position of the Earth, with reference to fixed stars, is nearly 5013 " west than that of previous year. This westward motion of the zodiac by 5013" a year, causes a slight motion in the case of the equinoxes also. This motion of the equinox is known as "precession of the equinoxes."

PRECESSIONAL DISTANCE *(Ayanamsa)*
Ayanamsa : Definitions

1)   *Ayanamsa* is the distance (in degrees, minutes and seconds) between the First Point (starting point) of the two zodiacs, namely, the Sidereal and Tropical Zodiacs.

2)   *Ayanamsa* is the difference between the the longitudes of the planets calculated for the Tropical and Sidereal Zodiacs.
3)   *Ayanamsa* is the distance between the Indian first point and vernal equinox, measured at an epoch.

The First Point of Aries of the Tropical Zodiac (along with the zodiac) is moving westward from the Sidereal Zodiac (and its First Point) at the rate of 5013" a year.

The *ayanamsa* is worked out based on the year of coincidence, ie.,   the year when the two zodiacs (and their first points) coincided.

THE YEAR OF COINCIDENCE

As regards the correct year of coincidence which is in effect the year of zero ayanamasa, there is some difference of opinion. Shri N. C. Lahiri has taken 285 AD as the year of coincidence where as for Dr B.V.Raman, it is 397 AD. There are some other astrologers like Shri Krishnamoorthy, in whose opinion, both of these years are wrong. Some young enthusiastic astrologers of modern times have also found out their ownayanamsas, rejecting all the above opinions!

DIFFERENT VALUES OF AYANAMSA

1.   Chitra paksha Ayanamsa. This is also called Lahiri's Ayanamsa. In this system the imaginary starting point of the nirayana zodiac is a point which lies at 180o from Chitra nakshatra and the year of coincidence of the first points of sayanana and nirayana zodiacs was 285 AD.

2.   Raman's Ayanamsa. This was the Ayanamsa followed by late Dr. B.V.Raman,the world famous Indian astrologer, founder President of ICAS and the Editor of Astrological Magazine.   In this system the year of coincidence is 397 AD. This Ayanamsa is 1o - 7' less than Lahiri's Ayanamsa.   The Astrological Magazine and some astrologers in Karnataka and Andhra Pradesh follow this Ayanamsa.

3. Krishnamoorthy's Ayanamsa. The Ayanamsa followed by Shri K.S.Krishnamoorthy, founder of Krishnamoorthy Paddhati. This is 6' less than Chitra paksha. The year of coincidence is 291 AD.

4. Revati paksha Ayanamsa. In this system Revati nakshatra is the point of 0o.It is 3o - 59' less than the Chitra paksha.

6. Driksiddha Ayanamsa. Jyotishacharya Late Sripada Venkata Ramana Daivajna Sarma's "Ayanamsa tatwa vivek" is a book on this subject.

5. Chandrahari's Ayanamsa. A young engineer from Kerala (b. 1961) is the discoverer and propagandist of this Ayanamsa. According to him Brahma was born in Rohini nakshtra and the Ayanamsa at that time, ie.,the commencement of this Kaliyuga was 46o - 40'.

THE VEXED PROBLEM OF AYANAMSA

As a result of the different opinions regarding the year of coincidence of 0° of the Moving and Fixed zodiacs different *ayanamsas*, known as 'Lahiri's Ayanamsa' 'Raman's Ayanamsa', 'Krishnamoorthy's Ayanamsa' etc., are in vogue. But the truth is that the exact date when both the zodiacs, or the first points of the fixed and movable zodiacs, coincided is definitely not known and hence the precessional distance, ie., *ayanamsa* varies from 19o to 23°. This is one of the sympathetic confusions of the modern Indian astrology.

Eg:- The *ayanamsa* on 1-1-2004 according to the different systems:

1.	Sayana (Modern Astronomy) :	=	0°	0'	0"
2.	Lahiri's Ephemeris	=	23° - 54' - 35"		
3.	Raman's Ayanamsa (Astrological Magazine)	=	22° - 27' - 50"		
4.	Kerala Pnchangas	=	23° - 54' - 35"		

However, the Government of India has adopted Lahiri's (or *Chitra paksha*) *ayanamsa* for the purpose of the Indian Calendar and Rashtriya Panchanga. It is the official *ayanamsa* and hence more popular.

METHOD TO WORK OUT AYANAMSA

The method to workout the *ayanamsa* is to subtract the year of coincidence from the year for which the *ayanamsa* is required and to multiply the result by 5013".

Eg: Raman's *Ayanama* for 1990:

Year for example	=	1990
Year of coincideence	=	397
Difference	=	1593
Motion of the zodiac per year	=	$50\,^{1}/_{3}$ "
Motion in 1593 years	=	1593 multiplied by
		$1593 \times \dfrac{151}{3} = 80181$ "

By converting the seconds into degrees and minutes, we get the Ayanamsa for 1990 as 22° - 16' - 21". Raman'a ayanamsa as on 1-1-1990 is 22-16-20. The diiference of 1" is due to the fact that the exact date of coincidence is not January 1st.

PURPOSE AND USE OF AYANAMSA

Indian astrology has adopted the *nirayana* positions of the planets, for all predictive purposes, which is to be arrived at after calculating the positions according to the *sayana* system and subtracting the *ayanamsa* from it. In other words, Indian astrological calculations are at first based on the tropical zodiac and then converted to *nirayana*. Since the ephemerises like Lahiri's Ephemeris etc. and the standard *panchangas* contain *nirayana* longitudes of planets, the problem of conversion of *sayana* longitudes into *nirayana* does not affect the students., at this stage.

THE SAYANA AND NIRAYANA SYSTEMS

The system of astronomy/astrology which follows the moving zodiac is called "Sayana" and the system which follows the fixed zodiac is "Nirayana".

SAYANA AND NIRAYANA LONGITUDES OF PLANETS

1.   What are Sayana and Nirayana longitudes ?   Which system of calculation do you prefer and why ?

2.   Difference between Sayana and Nirayana longitudes.

The longitude of planets based on sayana system is "sayana longitude" and the longitudes as per nirayana system is "nirayana longitudes". In other words, the sayana longitudes include ayanamsa and the nirayana system is without ayanamsa

LESSON 3
SIDEREAL AND TROPICAL ZODIACS

1.   Zodiac: Short Notes.
2.   Short note: Sidereal and Tropical zodiac.
3.   Explain in detail with diagram the concept of the Zodiac with its signs, names and symbolism, and constellational divisions.
4.   Explain the term Zodiac.
5.   What is the difference between Sidereal and Tropical Zodiacs ?
     Which do you prefer and why ?
6.   Highlight the basic diferences between the "Sayana" and "Nirayana" systems of
     measurement of Zodiac.
7.   Write short notes on Tropical Zodiac.
8.   State the differences betwen the sidereal and tropical zodiacs.
9.   Do you prefer the sidereal or the tropical zodiac? State your reasons.
10. Offer a defence of the sidereal zodiac.
11. Distinguish between the sidereal and the tropical zodiacs.
12. Explain the concept of Zodiac with a diagram?
13. Write the names of 12 (twelve) signs and 27 (twenty seven) constellations.

Angular measurements:

60"	(60 seconds, vikala)	=	1' (1 Minute, Kala)
60'	(60 Minutes, Kala)	=	1° (1 degree, Bhaga)
30°	(30 degree, Bhaga)	=	1s (1 sign, Rasi)
12s	(12 sign, Rasi	=	1 Zodiac, Rasi chakra

"It is a broad band or belt in the heavens extending 9 degrees on either side of the ecliptic, and known to the Hindus as Bhachakra or the Circle of Light. It is a circle and as such it knows no beginning or end. In order to measure the distance, an arbitrary point is established, which is called the first point of Aries. The zodiac revolves once in a day on its axis, from east to west."
--Dr. B.V. Raman, A Manual of Hindu Astrology

So the Zodiac is an imaginary road in the sky, stretching about 8° north and 9° south of the apparent path of the Sun in the sky, within which the planets move around the Earth, once in a day, from east to west.

The zodiac is divided into twelve equal parts called Signs (*Rasis*) (30° x 12 = 360°). These twelve divisions are called Aries (*Mesha*), Taurus (*Vrisha*) etc.

Signs of the Zodiac

1.	0° - 30°	Aries	*Mesha*	Medam
2.	30 - 60	Taurus	*Vrishabha*	Edavam
3.	60 - 90	Gemini	*Mithuna*	Mithunam
4.	90 - 120	Cancer	*Kataka*	Karkatakam
5.	120 - 150	Leo	*Simha*	Chingam
6	150 - 180	Virgo	*Kanya*	Kanni
7.	180 - 210	Libra	*Thula*	Thulam
8.	210 - 240	Scorpio	*Vrischika*	Vrischikam
9.	240- 270	Sagittarius	*Dhanus*	Dhanu
10.	270 - 300	Capricorn	*Makara*	Makaram
11.	300 - 330	Aquarius	*Kumbha*	Kumbham
12.	330 - 360	Pisces	*Meena*	Meenam

There are two zodiacs: one moving (Tropical) and the other fixed (Sidereal).

In the case of the Tropical Zodiac, each of these signs formerly occupied the similarly named constellations (groups of stars), but now, by precession of the equinoxes, they coincide with the constellations that bears the names of the preceding signs, that is, the constellation Pisces is now in Aries Sign.

The zodiac (*rasi chakra*) which reckons the Aries 0^0 (imaginary starting point of the zodiac) from the equinoctial point and which has precession, is the 'movable', 'tropical' or 'sayana' zodiac. Modern astronomy follows this zodiac.

But the zodiac considered in India for astrological purposes is a fixed one. In this case, the first point or 0^0 is based on a fixed star. In the Chitra paksha ayanamsa, it is a point between Revati and Aswini nakshatras, ie., 180^0 from Chitra nakshatra. So the zodiac considered in India for astological purposes is 'fixed', 'sidereal' or 'nirayana'.

This book contains: 1) Introduction (General information about Indian astrology). 2) Astronomy relevant to Astrology 3) Mathematical Astrology 4) Predictive Astrology, prepared with reference to the original classical texts, Brihat Jataka, Phaladeepika and Brihat Parasara Hora Sastra.

LESSON 4
ZODIAC AND CONSTELLATIONS

1. Short Notes : Constellations
2. How are the 27 nakshatras (constellations) related to the zodiac of the 12 signs ?
3. Show how the constellations are divided into the twelve signs. State whether the constellations reveal anything about the nature of the Zodiacal signs.
3. Are the Zodiacal signs related to the constellations ? Or are they arbitrary divisions ? Give reasons.
4. What is the Constellational Zodiac ? Does it differ from Sidereal and Tropical Zodiacs ? Give reasons ?
5. Differentiate between Tropical Zodiac and Constellation Zodiac?
   What are the divisions of Zodiac.

Though the *'Nakshatras'* in Indian asrtronomy and astrology are often called constellations, they are quite different from the actual constellations (groups of stars) in modern astronomy. In fact, *nakshatra* is also another system of division of the zodiac. In this system, the zodiac is divided into 27 equal parts and given the names of the constellations (group of stars) which are prominent within each division. While the zodiac with signs-divisions is called 'Zodiac of Signs'

(Rasi chakra), the zodiac with *nakshatra*-division is called 'Constellational Zodiac' or *Nakshatra chakra.*

S.No.	From		To	Nakshatra	Malayalam
1.	0° -	00'	13°-20'	Aswini	Aswati
2	13 -	20	26-40	Bharani	Bharani
3.	26 -	40	40-00	Krittika	Kartika
4.	40 -	00	53-20	Rohini	Rohini
5.	53 -	20	66-40	Mrigasira	Makayiram
6	66 -	40	80-00	Ardra	Thiruvatira
7.	80 -	00	93-20	Punarvasu	Punartam
8	93 -	20	106-40	Pushya	Pooyam
9.	106 -	40	120-00	Aslesha	Ayiliam
10.	120 -	00	133-20	Magha	Makam
11.	133 -	20	146-40	Purva Phalguni	Pooram
12.	146 -	40	160-00	Uttara Phalguni	Utram
13.	160 -	00	173-20	Hasta	Atham
14.	173 -	20	186-40	Chitra	Chitra
15.	186 -	40	200-00	Swati	Choti
16.	200 -	00	213-20	Visakha	Visakham
17.	213 -	20	226-40	Anuradha	Anizham
18.	226 -	40	240-00	Jyeshta	Trikketta
19.	240 -	00	253-20	Mula	Moolam
20.	253 -	20	266-40	Purvashadha	Pooradam
21.	266 -	40	280-00	Uttarashadha	Utradam
22.	280 -	00	293-20	Sravana	Tiruvonam
23.	293 -	20	306-40	Dhanishta	Avittam
24.	306 -	40	320-00	Satabhisha	Chatayam
25.	320 -	00	333-20	Purvabhadrapada	Pooruruttati
26.	333 -	20	346-40	Uttarabhadrapada	Utrattati
27.	346 -	40	360-00	Revati	

RASIS Vs. NAKSHATRAS

Rasis are twelve equal divisions of the zodiac. (30° x 12 = 360°). Similarly, *Nakshatras* are twenty seven equal divisions of the zodiac (13° - 20' x 27 = 360°). The Nakshatras have been divided into four parts called *paadas* (3°

- 20' x 4 = 13° - 20') and hence $2^1/_4$ *nakshatras* or 9 *nakshatra padas* form a Rasi.

JANMA RASI AND JANMA NAKSHATRA
Find out the longitude of the Moon at the time of birth and enter to the above table. You will get the following information from the table:-

JANMA RASI
Find from the table in which Rasi the longitude of the Moon falls. It is the Janma Rasi.

JANMA NAKSHATRA
1. What is Janma nakshatra ?

Find from the table the Nakshatra in which the longitude of the Moon Falls. It is the Janma Nakshatra *(Pirannal)*.

LESSON 5
EARTH AND THE CELESTIAL SPHERE
MEASUREMENTS AND CONNECTED TERMS

MEASUREMENTS:

60" (seconds, vikala)	=	1' (minute, kala)
60' (minute, kala)	=	1^{O} (degree, bhaga)
30° (degree, bhaga)	=	1s (sign, rashi)
12s (sign, rashi)	=	Zodiac (Rashi chakra)

The sky which is visible from the Earth is actually that part of the Universe which enfolds this planet.   The groups of stars which appear like luminous islands in the sky are Galaxies.   There are more than ten thousand crores of galaxies which contain crores and crores of stars. The Milky Way Galaxy *(Ksheera patham, Akasa Ganga)* is one among them.   The Sun is a star in that galaxy.   The observations with reference to the Earth is called Geocentric and the observations with reference to the Sun is Heliocentric. The Earth is a planet of the Sun. It rotates on its own axis in twenty four hours, resulting in days and nights and goes round the Sun in one year of $365\frac{1}{4}$ days , resulting in seasons.

EQUATOR (East-West-East)
(Bhoo madhya rekha, niraksha rekha, vishuvam)

Equator is an imaginary line round the Earth at equal distances from the north and south poles. It divides the Earth into northern and southern hemispheres. Geographically, equator is marked as 0^0 latitude. The north pole is 90^0 north and the south pole is 90^0 south from the equator

.

MERIDIAN *(Uchcha rekha)*
Meridian is an imaginary circle round the Earth, passing through a given place and the north and south poles. The meridian passing through Greenwich is considered as 0^0 longitude and it is called the Standard Meridian for the Earth. The Standard Meridian of India is having the longitude of 82^0-30' E. Meridian corresponds to the geographical longitude of a place. The M.C. (Medium Coeli) is the point of intersection of the ecliptic with the meridian of the place. The longitude of this point is measured along the ecliptic from the 'Firtst Point of Aries'.

LATITUDE and LONGITUDE
1. Short notes on Latitude and Longitude.
2. State with illustrtions the importance of latitude and longitude.
3. Short notes: Longitude
2. Define : Geographical and Celestial Latitude and Longitude.
4. Short note on Geographical longitudes (Rekhamsha) and latitudes (*Akshamsa*).

LATITUDE (*Akshamsa*: north--south)
Latitude of a place is its distance from the eqator, on its own meridian. It is reckoned in degrees from 0^0 to 90^0, northwards or southwards. 'S' or 'N' is also given to show where the place lies, ie., whether it is in the south hemisphere or north hemisphere. (Eg:- Latitude of Delhi is 28^0-39' N and Madras 13^0-04' N). The latitude of a place can be found out with the help of a Map or Atlas. It is also available in the books like Tables of Ascendants.

LONGITUDE (*Rekhamsa*: east--west)

The longitude of a place is its distance from 0^0 meridian (Greenwich) measured as an angle. It is expressed in degrees. ' E ' or 'W' is also indicated to show whether the place lies east or west of the meridian. (Longitude of Delhi 77^0-13' E. Madras 80^0-15' E.) The longitude is also reckoned in time at the rate of 24 hours for 360 degrees or 4 minutes per degree. Indian Standard Time is 5 hours 30 minutes ahead of Greenwich Time, because the longitude of Greenwich is 0^0 and Indian Standard meridian is 82^0-30' E. (82^0-30' x 4 = 330 minutes, ie., 5h 30mts.)

THE CELESTIAL SPHERE
(Khagola, nabhomandala)

The celestial sphere is an imaginary sphere in the space, surrounding the Earth. Celestial equator is a great circle of the celestial sphere, marked out by the extension of the Earth's equator.

CELESTIAL EQUATOR

Celestial equator is a great circle of the celestial sphere, marked out by he extension of the Earth's equator.

CELESTIAL LATITUDE AND LONGITUDE.

Celestial latitude (*kshepa*) is the angular distance of a heavenly body from the Ecliptic. Celestial longitude (*krantiamsa*) of a heavenly body is the angular distance of it measured along the Ecliptic, from the zero point (First point of Aries).

LATITUDE AND LONGITUDE OF PLANETS

It is the celestial latitude and longiude.

SAYANA AND NIRAYANA LONGITUDES
   1. What are Sayana and Nirayana longitudes? Which system of calculation do you prefer and why?
   2. Difference between Sayana and Nirayana longitudes.

The *sayana* longitude is as per the movable or tropical system of zodiac or longitude with *ayanamsa* and the *nirayana* longitude is as per fixed or sidereal zodiac or longitude without ayanamsa. You can get nirayana longitudes from sayana longitudes by subtracting the *ayanamsa*.Indian astrologers mainly follow *Nirayana* system ans the westerners follow *Sayana*.

LESSON 6

THE SUN AND THE STARS

THE UNIVERSE

The sky which is visible from the Earth is actually that part of the Universe that enfolds this planet. The groups of stars which appear like luminous islands in the sky are Galaxies. There are more than ten thousand crores of galaxies which contain crores and crores of stars. The Milky Way Galaxy (*Ksheera patham, Akasa Ganga*) is one among them. The Sun is a Star in that galaxy. The observations with reference to the centre of the Earth is called Geocentric and the observations with reference to the centre of the Sun is Heliocentric.

MILKYWAY GALAXY

The sun is an ordinary star in Milky way Galaxy. The galaxy is shaped like a concave lens. From left end to right end, it is about 98000 light years.Radio astronomy reveals that our galaxy is a spiral in constant rotation. Like our galaxy, there are crores of other galaxies, like Andomeda in the northern hemisphere.

NEBULAE

Nebulae are celestial matter which looks like clouds.   They are found in the galaxies. The local ones are galactic and those of other galaxies are called extragalactic nebulae.

THE SUN

The earth and some other heavenly bodies go round the sun.   These are planets, their satellites, comets, asteroids etc.   All these bodies with the sun is called the solar system.The sun gives us light and heat which are essential for living, for without heat and light there would be neither plant or animal life.   The sun is an enormously sized ball where there are hot and burning gases. The diameter of the sun is 8,64,000 miles   and at a mean distance of 93 million miles away from the earth. Since it is not a rigid body, the period of rotation about its equator is about 25 days and as much as 34 days about its poles.   Its visible surface is called the photosphere, having a temperature of 6000^{0} C. Its interior may have a temperature of 20 million degree centigrade.   The source of solar energy is due the "combustion" of hydrogen and its transformation into helium, resulting in the prouction of atomic energy in large quantities.   The nuclear reactions are responsible for the creation and maintenance of the solar energy.

THE STARS

The stars are huge bodies,   sometimes of the size of ten or twenty suns put together. They are burning balls of fire, containing chiefly hydrogen, iron, sodium, magnesium and calcium.   They are many hundreds of light years distance from us.   For example,   the star Arcturus is 38 light years away from us. The light from this star, we see to-day, has started 38 years ago, ie., how it appeared 38 years ago.

The stars twinkle where as the planets are steadily shining objects.   The light from the star is a single pencil of rays,   coming from one direction.   As it passes through the different layers of atmosphere varying in density, the eye sees them twinkling, while the flickering   in the case of the planets are counteracted by the light from some other direction of the planets, besides they reflect the light from the Sun.

The stars are of different brightness.   The stars just visible to the naked eye are of sixt magnitude of visual brightness. A star of fifth magnitude is $2^{1}/_{2}$ times brighter than that of the sixt magnitude.

There are many types of stars. A single star to the naked eye, when observed throuh a telescope, will be found to be really two stars.These types are called Double Stars or Binaries. There may be two stars   at different distances but in the same direction. These are called Optical Double or Visual Binaries.   But there are True Binaries which are two stars at the same distance from us, physically joined and revolving about a common axis.   Sirius is one such true binary.

The stars move in the firmament.   If the positions of the stars recorded between a gap of twenty or thirty years it will show that they have changed their positions , ie., the stars have proper motions.

The stars are also classified as red, yellow, white or blue according to their temperatures. Our sun is a yellow star.

Serious students may please read the famous book "A brief history of Time" by Prof. Stephen Hawking.
{Curtsey: Mysore University Teaching Notes on Astronomy.}

"Most of us are utterly oblivious to the fact that we are here for a very brief period and are an insignificant part of a vast and apparently unlimited universe. An insect crawling on the Himalayas has a comparatively greater significance   from the purely physical point of view.... ... yet it does not occur to many people to ask the very pertinent questions as to where we have come from, where we are going, and why we are here. " --I.K.Taimni

LESSON 7

SOLAR SYSTEM

1.   Briefly describe our solar system and the order in which the planets appear in the sky.

SOLAR SYSTEM   *(Soura yoodha)*

Some heavenly bodies like the Earth go round the Sun. These are planets, their satellites, asteroids and comets. All these bodies with the Sun is called Solar System.

PLANETS (*Grahas*)

The planets *rotate* about themselves and *revolve* round the Sun along their own paths called orbits. As per *Bode's Law*, there is a relation in the distances of the planets from Sun.

Planet	Add		Total	Divide by 10
Mercury	0	4	4	0.4
Venus	3	4	7	0.7
Earth	6	4	10	1.0
Mars	12	4	16	1.6
...	24	4	28	2.8
Jupiter	48	4	52	5.2
Saturn	96	4	100	10.0
Uranus	192	4	196	19.6

If the distance between the Sun and the Earth is taken as 1 Astronomical Unit, then Mercury is at 0.4 AU and Uranus at 19.6 AU. The important discovery made, using this theory, is that at 2.8 AU, there is no planet. But it was later found out that once there was a big planet there and it somehow burst out and became asteroids.

SATELLITES (*Upagraha*)

Satellites are heavenly bodies moving round the planets. The Moon is the satellite of the Earth. The Mars has 2 satellites, Jupiter 16, Saturn 17, Uranus 5, Neptune 4. There can be more sattellites yet to be discovered.

ASTEROIDS (*Chhinna graha*)

The asteroids are planetary fragments. They are also known as minor planets. There is a large number of asteroids between Mars and Jupiter. It is presumed that once there was a huge planet between Mars and Jupiter which was broken into pieces and these pieces occupying the orbit of that planet.

METEORS (*Ulkka*)

A meteor is a small body, rushing from the outer space to the atmosphere of the Earth, and becoming bright.

COMETS *(Dhooma ketu)*
1. Write short notes.
2. What are comets ? What is their origin? When do they appear ? Name any two and indicate their influence.
3. What are Comets ? Are they considered for Astrological predictions? What is the significance of the appearance of Comet 'Hyakuta' now orbiting earth ?
4. What is the effect of comets ? Explain.
5. What are comets and meteorites.
6. Explain comets.

A comet which consists of a 'coma' (a shining nucleus) and unseen hydrogen clouds, is a star-like heavenly body, with a tail of light. It moves across the sky through regular routes. The tail which appears only when coming near the Sun, always keeps to the opposite direction to the Sun. More than 600 comets have since been identified and indexed.

The comets whose motion can be calculated and the date of their return predicted are called periodic comets. Halley's Comet appears at an interval of 76 years. It last appeared in 1986. Encle's Comet appears at an interval of 3.3 years. Non-periodic comets are more in number than the periodics.

LESSON 8
INNER AND OUTER PLANETS

1. Define Inner and Outer planets of Solar System. Draw a neat diagram of Solar System.

2. Which are the extra-Saturnine planets ? Show the order in which they appear in the sky.

INNER AND OUTER PLANETS

The planets whose orbits lie between the Earth and the Sun are called inner, interior or inferior planets. (Mercury and Venus).

The planets whose orbits lie outside the orbit of the Earth are called outer, exterior or superior planets (Mars, Juoiter, Saturn etc.).

The planets whose orbits lie outside the orbit of Saturn are called extra-Saturnine planets. (Uranus, Neptune and Pluto).

LESSON 9

PLANETS CONSIDERED IN INDIAN ATROLOGY FOR PREDICTIVE PURPOSES

The planets considered in Indian Astronomy for astrological purposes are called *Grahas*. The *grahas* are:

1. *Surya* - The Sun
2. *Chandra* - The Moon

3.   *Kuja* - Mars
4.   *Budha* - Mercury
5.   *Guru* - Jupiter
6.   *Sukra* - Venus
7.   *Sani* - Saturn
8.   *Rahu* (Moon's Ascending Node); and
9.   *Ketu* (Descending Node)

.

The concept of *grahas* in Indian astronomy and astrology differs from that of a 'planet' in western system.

In Indian system, the Sun (a star), the Moon (sattellite of the Earth) and the two Nodes (Rahu and Ketu) are condidered as *grahas*.

In Kerala, *Gulika (Maandi),* an astrological *Upagraha*, has also been given the status of a *graha* for all astrological purposes, especially in *prasna* matters.

EXTRA-SATURNINE PLANETS

1. Should we consider extra-Saturnine planets like Uranus, Neptune and Pluto ?
Give reasons.

The extra-Saturnine planets   like Uranus, Neptune and Pluto have no place in Indian   astronomy and astrology.   But some enthusiastic astrologers include them also. According us, only the nine planets discussed classical texts need be considered for predictive purposes.

LESSON 10
PLANETS AND WEEK DAYS

RATE OF MOVEMENT

Planet	Speed per day
Moon	$13^{o} - 10' - 34''$
Mercury	$4^{o} - 5' - 32''$

Venus	1^O - 36' - 7"	
Mars	0° - 31' - 26"	
Jupiter		0° - 4' - 59"
Saturn	0° - 2' - 0"	

The velocity of the planets is related to the distance from the Sun. 'Apparent place' of any celestial object is the position at which the celestial object would actually be seen from the centre of the Earth, displaced by planetary aberration and refers to the true equinox and equator.

PLANETS AND WEEK-DAYS

1. Do you see any relation between order of the seven major planets of the Indian system and our week days ?
2. How week days are co-related to time measures?
3. Short note: Names of the days of week and rationale behind the scheme.
4. How do the days of the week get their names in Hindu Astrology?

The Indian astronomical / astrological day is from sunrise to the next sunrise and it is divided into 24 *Horas* (*Hora* is equal to one hour). Following the principles in week days, the lordship of the *horas* have been allotted to the seven major planets. The first *hora* starts at the sunrise and the day will be in the name of the planet who owns the first hora.

Order of the week-days is related to the system of division of horas. The lordship of the horas by the seven planets is based on their velocity in revolotion around the Sun.

S. No. Planet	Rate of speed per day		
1. Saturn	0°	2'	0"
2. Jupiter	0	4	59
3. Mars	0	31	26
4. Sun (Earth)	0	59	8
5. Venus	1	36	7
6. Mercury	4	5	32
7. Moon	13	10	34

The horas repeat in this manner and the 25th hora becomes first hora of the next day. The table of *kala horas* and the week-days based on them is given below. The authority for this arrangement is Suryasiddhanta.

Hora	Sunday	Monday	Tuesday	Wednesday	Thursday	Friday	Saturday
1	Ravi	Chandra	Kuja	Budha	Guru	Sukra	Sani
2	Sukra	Sani	Ravi	Chandra	Kuja	Budha	Guru
3	Budha	Guru	Sukra	Sani	Ravi	Chandra	Kuja
4	Chandra	Kuja	Budha	Guru	Venus	Sani	Ravi
5	Sani	Ravi	Chandra	Kuja	Budha	Guru	Sukra
6	Guru	Sukra	sani	Ravi	Chandra	Kuja	Budha
7	Kuja	Budha	Guru	Sukra	Sani	Ravi	Chandra
8	Ravi	Chandra	Kuja	Budha	Guru	Sukra	Sani
9	Sukra	Sani	Ravi	Chandra	Kuja	Budha	Guru
10	Budha	Guru	Sukra	Sani	Ravi	Chandra	Kuja
11	Chandra	Kuja	Budha	Guru	Venus	Sani	Ravi
12	Sani	Ravi	Chandra	Kuja	Budha	Guru	Sukra
13	Guru	Sukra	sani	Ravi	Chandra	Kuja	Budha
14	Kuja	Budha	Guru	Sukra	Sani	Ravi	Chandra
15	Ravi	Chandra	Kuja	Budha	Guru	Sukra	Sani
16	Sukra	Sani	Ravi	Chandra	Kuja	Budha	Guru
17	Budha	Guru	Sukra	Sani	Ravi	Chandra	Kuja
18	Chandra	Kuja	Budha	Guru	Venus	Sani	Ravi
19	Sani	Ravi	Chandra	Kuja	Budha	Guru	Sukra
20	Guru	Sukra	sani	Ravi	Chandra	Kuja	Budha
21	Kuja	Budha	Guru	Sukra	Sani	Ravi	Chandra
22	Ravi	Chandra	Kuja	Budha	Guru	Sukra	Sani
23	Sukra	Sani	Ravi	Chandra	Kuja	Budha	Guru
24	Budha	Guru	Sukra	Sani	Ravi	Chandra	Kuja
25	Chandra	Kuja	Budha	Guru	Venus	Sani	Ravi

EXAMPLE:
Please check the correctness of the table with reference to the week-days of any week in your Ephemeris/Panchanga. Using the table, you can also find out the *hora* of a particular day and time.

LESSON 11
RAHU AND KETU

1. Offer a brief note on Rahu and Ketu. Do you think they are planets?
2   Why are Rahu and Ketu considered as "Chhaya Grahas" ?   Do they have dual motions (Direct and Retrogression) and True position ?

THE NODES (*Sandhi*, joint)
The nodes are the two points on the celestial sphere at which the plane of an orbit of any rotating celestial object intersects a reference plane.   In respect of the Moon, these two points are known as Ascending Node and Descending Node (*Rahu* and *Ketu*).

RAHU AND KETU

Rahu and Ketu are two nodal points where the path of the Moon cuts the path of the Earth. When the Moon crosses the ecliptic, in the course of going from south to north, it is Rahu (Ascending Node). The latitude of the Moon at this point is zero and it is on the increase from negative to positive. Similarly when the Moon crosses the ecliptic, going from north to south, it is Ketu (the Ascending Node).

The Moon and the Sun get eclipsed at or near these points on Full Moon day (*Poornima*) and New Moon day (*Amavasya*) respectively.

The Nodes move anti-clockwise at a distance of 180 degrees. (Eg: When the longitude of Rahu is 1° , the longitude of Ketu is $180^{o.}$)

'True Rahu' is by considering the actual oval / elliptical shape of the orbits of the Moon and the ecliptic, whereas 'Mean Rahu' is calculated by considering these as perfect circles. The greatest difference between thetwo is 1^3_4 degrees only. As Mean Rahu is not factully correct, True Rahu is generally considered for astrological purposes.

LESSON 12
MANDI

    1. Shortnote: Mandi.
    2. How will you calculate Gulika / Mandi in a birth horoscope.

Mandi or *Gulika* is an astrological *Upagraha*. In Kerala it has been given the status of the tenth planet and its position is recorded in the horoscope just like the other nine major planets. In *'Prasna'*, *Gulika* is given more importance than the other nine planets and before the longitude of planets are worked out, the position of *Gulika* is calculated and recorded in *Prasna-chart.*

.

    In determination of *Muhurta* also, *Gulika* has a very important role. *Gulika-kalam* is treated as a very inauspiscious time even by the public who do not know astrology. In determining the time of death, rectification of birth-time etc. also the *Gulika* is considered.

Gulika rises twice a day, that is, once in the day-time and once in the night, at fixed times. On Sundays, it rises at 26th *ghati* after sunrise and 10th *ghati* after sunset. The time reduces by four *ghatis* daily and the rising time in night becomes the rising time of the fifth day, day-time.(See Table).

GULIKODAYA

Day	After sunrise *(Ghati)*	After sunset *(Ghati)*
Sunday	26	10
Monday	22	6
Tuesday	18	2
Wednesday	14	26
Thursday	10	22
Friday	6	18
Saturday	2	14

The method to find out the lngitude of Mandi at the time of birth will be discussed Mathematical Astrology.

Some astrologers are of the opinion that Maandi and Gulika are not the same planetbut two different planets. But that point is not relevant here. We may, for the time being, treat Maandi and Gulika as one and the same Upagraha, as is relevant for our study purposes, and let the scholars ponder over the issue.

LESSON 13
OTHER UPAGRAHAS

3. Explain how to calculate Upagrahas. What do they indicate ?

The other *Upagrahas* are secondary planets in Indian astrology. They are not physical bodies but mathematically computed astronomical points on the ecliptic with reference to the Sun's longitude. The important *Upagrahas* are : -

1. Sun's *nirayana* longitude + 133°-20'
   (or 10 nakshatras: 13-20 x 10) =   *Dhuma*
2. 360°-- *Dhuma* =   *Patha*
3. Patha + 180° =   *Paridhi*
4. 360° -- Paridhi =   *Indra chapa*
5. Indra chapa + 16°-40' =   *Sikhi*

There are some more *Upagrahas* like:
1. *Yamakantaka* (S/o Jupiter)
2. *Ardha prahara*   (S/o Mercury)
3. *Kala* (S/o Saturn)
4. *Kala* (S/o Mercury)
5. *Upaketu* (S/o Ketu

LESSON 14
RIGHT ASCENSION AND DECLINATION

In the case of the heavenly bodies, there are two different systems of co-ordinates:
a) celestial longitude & latitude: and
b) Right Ascension and Declination.

In the first system (celestial latitude and longitude) which is followed by the astrologers, the measurements are along and perpendicular to the ecliptic whereas with the second system (right ascension and declination), adopted by the astronomers, the measrements are along and perpendicular to the celestial equator.

CELESTIAL LATITUDE
Celestial latitude (*kshepa*) is the angular distance of a heavenly body from the Ecliptic.

CELESTIAL LONGITUDE

Celestial longitude (*krantiamsa*) of a heavenly body is the angular distance of it measured along the Ecliptic, from the zero point (First point of Aries). Eg: Longitude of planets.

DECLINATION

The 'Declination' is the angular distance of a heavenly body from the celestial equator. It is positive or negative according as the celestial object is situated within the

northern or southern hemisphere.

RIGHT ASCENSION

'Right Ascension' is the angular distance on the celestial sphere measured eastward along the celestial equator from the *Vernal Equinox*, to the *hour-circle* passing through the celestial object.

LESSON 15
MOUDYA AND *VAKRA*

MOUDYA (COMBUSTION)

The planets remain invisible to the naked eye for some days at the time of conjunction with the Sun. They become combust when coming very near to the Sun, that is, when they are within a particular distance from the Sun, eitherside:-

Planet	Distance either side of the Sun	Difference in opinion
Moon	12°	-
Mars	17°	14°
Mercury	14°	-
Jupiter	11°	12°
Venus	10°	-

Saturn	15°	16°

==

As regards Mars, Jupiter and Saturn (the outer planets) there is some difference of opinion as 14, 12, and 16 degrees respectively. The rule of combustion does not apply to the nodes (Rahu and Ketu).

VAKRA (RETROGRESSION)

Retrogresion is the backward motion of the planets. Actually the planets never go backward but due to the movements of the Earth and the other planets, occupying different angles, it so appears.

The inner planets become retrograde when they are in between the Earth and the Sun and the outer planets retrograde when the Earth is in between them and the Sun, ie., they are nearer to the Earth. In the case of the inner planets, their faster angular velocity than the Earth causes the apparent retrogression. In the case of the outer planets, it is the velocity of the Earth that causes the phenomena.

LESSON 16
THE LUNI-SOLAR YEAR

LUNAR DAYS, MONTHS AND YEARS

Lunar day is the day based on the movement on the Moon. *Tithi* is the Indian lunar day. The average time of a *tithi* is 23 hours 37 minutes 28 seconds.

A lunar month is 30 such *tithis*, starting from one new-moon (*Amavasi*) to the next new-moon. A lunar month is 29.5 mean solar days.

A lunar year is twelve such months (29.53 x 12) 354.36 days, against the solar year of 365.25 days. In other words, a lunar year is 11 days short of a solar year, as a lunar day is 22 minutes less than a solar day.

SOLAR DAYS, MONTHS AND YEARS

A solar day is the day based on the apparent movement of the Sun (or rotation of the Earth on its axis). The solar day, according to western system, is from midnight to midnight, but in Indian system, it is from one sunrise to next sunrise.

' *Savana* day' is the apparent solar day from sunrise to sunrise. It is longer than a sidereal day (*nakshatra divasa*). The duration of a solar day is 24 hours against the sidereal time of 23hrs 56 mts. ('Sidereal Time' will be explained in Mathematical Astrology.) *Savana* year is 360 Mean Solar Days.

'Mean solar day ' is average length of a day in a solar year. Solar month is the time taken by the Sun to move through a sign of the zodiac. A solar year is twelve solar months.

'Tropical year' is the time of passage by the Sun from one Vernal Equinox to the next Vernal equinox. The Sun appears to move about one degree of the zodiac in a day and accordingly travels one sign in about 30 days and the entire zodiac in 365.25 days. The time taken by the Sun to cover the distance of one sign is a solar month and the time taken for a full circuit of the zodiac is a solar year.

LUNI-SOLAR YEAR

1.   Explain the terms lunar year, solar year and luni-solar year.

The luni-solar year is the system of solar year with lunar months. The difference of 11 days (365 - 354) a year is adjusted by a thirteenth month once in 3 years. This thirteenth lunar month is called *Malamasa*.

There are three types of *Malamasas*, namely, *Amhaspati, Adhimasa* and *Samsarpa*:

1.   *Amhaspati* is a lunar month where in two *Surya Sankrantis* occur.
2.   *Adhimasa* is a lunar month which comes as an excess month (13th month) in every three years.
3.*Samsarpa* is the month between two *Surya Sankrantis*, where two

4.*Amavasyas* end in a solar month. It is the lunar month preceding *Amhaspati*. (*Surya samkranti* / *Surya samkramam* is the Sun's entering to a new *Rasi*.)

India followed solar year with lunar months and a synhronised luni-solar year since early Vedic period. To get exact correspondence in solar and lunar years, seven *adhimasas* were added in a period of 19 years.

BEGINNING OF THE YEARS

The solar year begins, as per the National calendar, on March 21 (Venral Equinox or Sayana Sun: 0^0·) All others except Kerala, entry of the Sun to Mesha Rasi (April 14). In Kerala, entry of the Sun to Simha Rasi. "Mesha vishu samkramam" is also an important day in Kerala but the Kollavarsham begins on Chingam1). As regards Lunar year, diferent systems are followed in the case of different eras.

LESSON 17
THE MEANING OF PANCHANGA

What are Panchangas ?
How are these calculated ?
Illustrate your answer with suitable examples.

PANCHANGAS

Nakshatra vaara tithaya: karanaani yogaa
Panchaangametadatharaasiyutam shadangam.. (Madhaviyam)

The '*Panchanga*' is an Indian almanac which consists of five *angas* (parts), namely,

     1) Nakshatra,
     2) Vara,
     3) Tithi,
     4) Karana ; and

5) *Nitya Yoga.*

In addition to these, modern *Panchangas* contain many more astronomical and astrological informations such as longitudes of planets, sunrise and sunset, eclipses, *rasi-pramana, muhurtas* etc.

1. Nakshatra

The *Nakshatra* is determined with rerefence to the movement of the Moon, through the 27 constellations of 13° -20' each (13-20 x 27 = 360°) of the zodiac. Eg: If the longitude of the Moon is in between 0° and 13° -20' ,   it is Aswini Nakshatra. Thererfore, to find out the Nakshatra of a particular day,   the longitude of the Moon at that time is to be divided by 13° -20'.   The *Janma Nakshatra* is the *Nakshatra* in which the Moon is posited at the time of birth.

S.No.	From	To	Nakshatra
1.	0° - 00'	13°-20'	Asvini
2	13 - 20	26-40	Bharani
3.	26 - 40	40-00	Krittika
4.	40 - 00	53-20	Rohini
5.	53 - 20	66-40	Mrigasirsha
6	66 - 40	80-00	Ardra
7.	80 - 00	93-20	Punarvasu
8	93 - 20	106-40	Pushya
9.	106 - 40	120-00	Aslesha
10.	120 - 00	133-20	Magha
11.	133 - 20	146-40	Purva Phalguni
12.	146 - 40	160-00	Uttara Phalguni
13.	160 - 00	173-20	Hasta
14.	173 - 20	186-40	Chitra
15.	186 - 40	200-00	Svati
16.	200 - 00	213-20	Visakha
17.	213 - 20	226-40	Anuradha
18.	226 - 40	240-00	Jyeshta
19.	240 - 00	253-20	Mula
20.	253 - 20	266-40	Purvashadha
21.	266 - 40	280-00	Uttarashadha
22.	280 - 00	293-20	Sravana

23.	293 - 20	306-40	Dhanishta (Sravishta)
24.	306 - 40	320-00	Satabhishaj
25.	320 - 00	333-20	Purvabhadrapada
26.	333 - 20	346-40	Uttarabhadrapada
27.	346 - 40	360-00	Revati

 Example: Find the *nakshatra* at 10 AM on 1-1-2004

As per Lahiri' Ephemeris (page 16), the longitude of the Moon at 5.30 AM on 1-1-2004 is 0-3-35. Since 0° to 13°-20' is Aswini, it is Aswini Nakshatra.

Cross check: As per page 59 of the Ephemeris, from1-09 AM on 1-1-2004 to 1-09 AM on 2-1-2004, the *nakshatra* is Aswini.

2. Vara

Days of the week, Sunday, Monday etc. Duration is from one sunrise to next sunrise.

3. Tithi

How Tithi is calculated ? Explain with example. 6/97

Tithi is Indian lunar day and it is based on the difference in the longitudes of the Moon and the Sun.

On all new moon days the Sun and the Moon will come to one point in the zodiac, ie., the longitudes of both the bodies will be one and the same. As the fast moving Moon overtakes the Sun and reaches at a point of 12 degrees ahead of the Sun, the first *tithi* or the first lunar day of the bright fortnight is over. In this way, each time when the Moon reaches another point (12 degrees), it is another *tithi*. This process continues till the full moon day which is 180° (12° x 15) from the Sun.

When the Moon moves away from this point (180°), the dark fortnight begins. In this manner, when the Moon completes another 15 *tithis* of 12° each and reaches again the same point at $360^o/0^o$, it is agin the new moon day.

1.BrightFortnight (Shukla Paksha)

0	12	Pratipada
12	24	Dvitiya
24	36	Tritiya
36	48	Chaturthi
48	60	Panchami
60	72	Shashti
72	84	Saptami
84	96	Ashtami
96	108	Navami
108	120	Dasami
120	132	Ekadasi
132	144	Dvadasi
144	156	Trayodasi
156	168	Chaturdasi
168	180	Pournami (Full Moon)

2.Dark Fort night   *(Krishna Paksha)*

180	192	Pratipada
192	204	Dvitiya
204	216	Tritiya
216	228	Chaturthi
228	240	Panchami
240	252	Shashti
252	264	Saptami
264	276	Ashtami
276	288	Navami
288	300	Dasami
300	312	Ekadasi
312	324	Dvadasi
324	336	Trayodasi
336	348	Chaturdasi
348	360	Amavasi (New Moon)

ADHIK TITHI AND KSHAYA TITHI

1. State what contributes to "Adhik Tithi" and "Kshay Tithi". 12/2002.

Due to various reasons the movement of the Moon is not at a fixed rate. It moves sometimes fast and sometimes slow and the speed varies from about 12° to about 15° in 24 hours. The variation in the movement of the Moon may sometimes result in losing or gaining of a *tithi*.

The *tithi* which is at the time of sunrise is the *tithi* for that day. The *tithi* which starts after sunrise and ends before the next sunrise is said to be missed in that fortnight. If a *tithi* which starts just before sunrise and ends after the next sunrise will be having two days in its name in that fortnight. The cases of two *tithis* on the same day or of no *tithi* in a particular day are called as *Adhik tithi* and *Kshaya tithi* respectively. (This system was introduced for day to day working of the society and hence will not be correct as per calculations in the following exmple cases.)

Method to find out the *Tithi*: $\dfrac{\text{Longitude of the Moon} \ -- \ \text{long. of the Sun,}}{12}$

Example: 1) Find the Tithi on 1-1-2004 at 10 AM:

Long. of the Moon at 5-30 AM = 0s-3°-35' = 3°-35'.
(As it is less than the Sun's longitude, 360° is to be added with it).

3°-35' + 360° $= 363^o$ -35'.

Long. of the Sun 8s-15°-58', that is, 255°-58'.

363°-35' -- 255°-58' $= 104^o$-37'.

As per the table, 96° - 108° is Suklapaksha Navami.   But by 6-22 AM, Moon will cross the border of 108°, resulting in the beginning of Dasami. As per the Ephemeris from 6-22 AM on 1-1-2004 to 8-52AM on 2-1-2004, it is *Suklapaksha Dasami.*

2)   Find the Tithi at 10 AM on 16-1-2004

Longitude of the Moon: 191-35 + 360	= 551-35
Longitude of the Sun:	271-15
Difference	280-20

From 276° to 288°, it is Krishna paksha Navami. The calculation is correct as per the Ephemeris also. From 21h-29m on 15-1-2004 to 19h-36m on 16-1-2004, the tithi is *Krishna paksha navami.*

Please note that the rule governing the "adhik tithi" and "kshaya tithi" has not been applied here. The tithi at the sunrise is considred as tithi of that day in certain parts of the country for social purposes, but the real tithis are as shown in the examples and also in the Lahiri's Ephemeris.

Similarly, in Kerala, the difference between the actual astronomicl position and the position for social purposes, is prevalent in the cases of *Janma nakshatras*, beginning of solar months etc. also. The *Pirannal* (birth day star) is the *nakshatra* which continues $2^1/_2$ hours after the <u>sunrise</u> of the place. The nakshatra for *sraddha* is the star whih has a balance of $2^1/_2$ hours before <u>sunset</u>. If the star comes twice in the month, ie., in the first week and the last week, the star of the first week is considred for *sraddha* purposes and the second is taken for *Janmadina* purposes.

4. Karana

1. Show how the Tithi, yoga and karana are calculated.
2. How are Yogas and Karanas calculated ?

A *Karana* is half a *tithi*. Therefore, there are two *karanas* in a *tithi*., ie., first part of the *tithi* (1 to 6 degrees) is one *karana* and and second part (7 to 12 degrees) another *karana*. Like *tithi*, *karana* is also decided with reference to the time taken by the Moon to gain over the Sun by 6 degrees, against12 degrees for a *tithi*. There are two types of *karanas*, that is, *Chara karanas* and *Sthira karanas*.

<u>Chara Karanas :</u>
1. *Bava (Simha)*
2. *Balava (Puli)*
3. *Kaulava(Panni)*
4. *Taitila (Kazhutha)*
5. *Garaja (Ana)*
6. *Vanij (Pasu)*
7. *Vishti*

<u>Sthira Karanas:</u>
1. *Sakuni (Pakshi - Pullu)*
2. *Chatushpada (Nalkkali)*
3. *Naga (Pambu)*

4. Kintugna (Puzhu)

Out of these, *chara karanas* repeat eight times, starting from second part of *Shukla Pratipada* (7 x 8 = 56). The *stira karanas* start from the first part of *Krishna Chaturdasi* and complete the circle of 60 *karanas* in 30 *tithis*.

Bright Fortnight

Tithi			Karana			
1.	Pratipada	0 - 6	Kintughna (1)	6 - 12	Bava (2)
2.	Dvitiya	12 - 18	Balava (3)	18 - 24	Kaulava (4)
3.	Tritiya	24 - 30	Taitila (5)	30 - 36	Gara (6)
4.	Chaturthi	36 - 42	Vanij (7)	42 - 48	Vishti (8)
5.	Panchami	48 - 54	Bava (9)	54 - 60	Balava (10)
6.	Shashti	60 - 66	Kaulava (11)	66 - 72	Taitilan (12)
7.	Saptami	72 - 78	Gara (13)	78 - 84	Vanij (14)
8.	Ashtami	84 - 90	Vishti (15)	90 - 96	Bava (16)
9.	Navami	96 - 102	Balava (17)	102 - 108	Kaulava(18)
10.	Dasami	108 - 114	Taitila (19)	114 - 120	Gara (20)
11.	Ekadasi	120 - 126	Vanij (21)	126 - 132	Vishti (22)
12.	Dvadasi	132 - 138	Bava (23)	138 - 144	Balava (24)
13.	Trayodasi	144 - 150	Kaulava (25)	150 - 156	Taitila (26)
14.	Chaturdasi	156 - 162	Gara (27)	162 - 168	Vanij (28)
15.	Pournami	168 - 174	Vishti (29)	174 - 180	Bava (30)

Dark Fortnight

16.	Pratipada	180 - 186	Balava (31)	186 - 192	Kaulava (32)
17.	Dvitiya	192 - 198	Taitila (33)	198 - 204	Gara (34)
18.	Tritiya	204 – 210	Vanij (35)	210 - 216	Vishti (36)
19	Chaturthi	216 - 222	Bava (37)	222 - 228	Balava (38)
20	Panchami	228 - 234	Kaulava (39)	234 - 240	Taitila (40)
21	Shashti	240 - 246	Gara (41)	246 - 252	Vanij (42)
22.	Saptami	252 - 258	Vishti (43)	258 - 264	Bava (44)
23	Ashtami	264 - 270	Balava (45)	270 - 276	Kaulava (46)
24	Navami	276 - 282	Taitila(47)	282 - 288	Gara (48)
25	Dasami	288 - 294	Vanij (49)	294 - 300	Vishti (50)

26	Ekadasi	300 - 306	Bava (51)	306 - 312	Balava (52)
27.	Dvadasi	312 - 318	Kaulava (53)	318 - 324	Taitila (54)
28.	Trayodasi	324 - 330	Gara (55)	330 - 336	Vanij (56)
29.	Chaturdasi	336 - 342	Vishti (57)	342 - 348	Sakuni (58)
30.	Amavasi	348 - 354	Chatushpada(59)	354 - 360	Naga (60)

Method to work out the *karanas*: Longitude of Moon -- longitude of the Sun, divided by 6.   Example: Find the Karana for 1-1-2004, **10 AM**.

      1-1-2004, **5.30**AM   Longitude of the Moon: 363 - 35
                                                   Longitude of the Sun: 255 - 58
                                                   Total   107 - 37.

107-37 divided by 6 = 17-57, ie., 18th Karana, ie., Kaulava.As per Lahiri's Ephemeris, at 6-22AM, Kaulava changes to Taitila. Therefore, the Karana at 10 AM on 1-1-2004 is Taitila.

5. Nityayoga

1.   Short notes: Yogas
2.   Show how the Tithi, yoga and karana are calculated.

Whereas the *tithi* is based on the differences between the longitudes of the Moon and the Sun, the *Nitya yogas* are based on the total of the longitudes of the Sun and the Moon.   There are 27 *Nitya yoga*s with reference to the 27 *nakshatras*, starting from *Pushya,* the 8th nakshatra (See table).   Therefore, $93^0 - 20'$ (the longitude upto the 7th nakshatra, Punarvasu, ie.,   13-20 x 7 = 91-140 = 93^0-20') is also to be added to the total of the longitudes of the Sun and the Moon.

The method to work out the *nitya yoga*: $\frac{\text{Long. of Sun + long of Moon + 93-20}}{13^0 \text{ -20'}}$

Example: Find the *nitya yoga* for 1-1 - 2004 at 10 AM.

Long. of the Moon at 5-30 AM   :   0s - 3^0-35'   =   3 - 35

Long. of the Sun at 5-30 AM : 8s -15⁰-58' = 255 - 58

Asvini to Punarvasu : 3s - 3⁰-20' = 93- 20

Total : 351⁰-113' = 352- 53

346-40 to 360 is Revati nakshatra.

From Pushya to Revati = 20 nakshatras. Hence 20th yoga from Vishkamba = Siva.

As per Lahiri's Ephemeris, it is 20th Yoga upto 6-39 PM.

No.	From	To	Nakshatra
1.	93 - 20	106 - 40	Pushya
2.	106 - 40	120 - 00	Aslesha
3.	120 - 00	133 - 20	Magha
4.	133 - 20	146 - 40	Purva Phalguni
5.	146 - 40	160 - 00	Uttara Phalguni
6.	160 - 00	173 - 20	Hasta
7.	173 - 20	186 - 40	Chitra
8.	186 - 40	200 - 00	Svati
9.	200 - 00	213 - 20	Visakha
10.	213 - 20	226 - 40	Anuradha
11.	226 - 40	240 - 00	Jyeshta
12.	240 - 00	253 - 20	Mula
13.	253 - 20	266 - 40	Purvashadha
14.	266 - 40	280 - 00	Uttarashadha
15.	280 - 00	293 - 20	Sravana
16.	293 - 20	306 - 40	Dhanishta
17.	306 - 40	320 - 00	Satabhishaj
18.	320 - 00	333 - 20	Purvabhadrapada
19.	333 - 20	346 - 40	Uttarabhadrapada
20.	346 - 40	360 - 00	Revati
21.	0 - 00'	13 - 20	Asvini
22	13 - 20	26 - 40	Bharani
23.	26 - 40	40 - 00	Krittika
24.	40 - 00	53 - 20	Rohini
25.	53 - 20	66 - 40	Mrigasirsha
26	66 - 40	80 - 00	Ardra
27.	80 - 00		

LESSON 18
PHASES OF THE MOON

1. Short Notes : Phases of the Moon.

We know that the Moon has no light of its own.   It reflects the light received from the Sun.   The side opposite to the direction from which the Sunrays reach the Moon will be dark. Hence there are two hemispheres of he Moon -- one bright and the other dark.

The Moon moves along its orbit around the Earth. When the Earth comes inbetween the Sun and the Moon, the bright part of the Moon is towards the Earth and hence the full disc of the Moon is visible to us. This is Purnima, Full Moon.

Similarly, when the dark part of the Moon is towards the Earth, the Moon is not visible from the Earth and it is Amavasya, New Moon.

LESSON 19
ECLIPSES

1. What are eclipses?
2. How are eclipses formed?
3. Short note : Solar eclipse. Lunar eclipse.
4. Define eclipse.
5. Write a note on solar and lunar eclipses

6. What is an eclipse? Discuss lunar eclipse with diagram.

      An eclipse is the total or partial cutting of the light of the Sun.   When the Moon comes in between the Earth and the Sun, it results in the eclipse of the Sun. Similarly, when the Earth comes in between the Moon and te Sun, it is the eclipse of the Moon.   The eclipse of the Sun and the Moon occur on new moon day (*Amavasya*) and full moon day (*Pournami*) respectively.   In other words, a lunar eclipse takes place when the Moon passes through the shadow of the Earth on Pournami and the solar eclipse will ocur when the Moon is in between the Earth and the Sun on Amavasya.   The Moon and the Sun get eclipsed on Poornima and Amavasya respectively at the nodal points (Rahu and Ketu).

When the whole of Moon's disc is obscured,   the eclipse is called a 'total eclipse' and when only a part of it is obscured it is 'partial eclipse'.

Solar eclipse are of three kinds: 1) Total eclipse. 2) Partial eclipse; and 3) Annular eclipse. The annular solar eclipse takes place when the Sun is nearest to the Earth and Moonfarthest and other conditions remain the same as that of total eclipse.

ASTRONOMICAL TERMINOLOGY
(Compiled from authentic books)

ALTITUDE
The altitude of a heavenly body is its distance from the horizon measured on the vertical, drawn through the body, from the zenith of the observer to the horizonal circle.

APHELION
Aphelion is the point on a planetary orbit when it is at the greatest distance from the Sun.

APOGEE
Apogee is the point at which a body in orbit around the Earth is at the greatest distance from the Earth.

APPARENT POSITION
Apparent place of any celestial object is the position at which the celestial object would actally be seen from the centre of the Earth.

ASCENDANT (*Lagna*)
The Ascendant is the point of intersection of the ecliptic, at the given time, with the eastern horizon of the place.   In astrology, it is the rising sign at the time of birth etc.

ASPECT
Aspect is the apparent position of any of the planets or the Moon, relative to the Sun, as observed from the Earth and angular relationship between two celestial objects. In astrology,   it is called *drishti* of the planets.

ASTEROIDS *(Chhinna graha)*
The asteroids are planetary fragments. They are also known as minor planets.   There is a large number of asteroids between Mars and Jupiter.

ASTRONOMY *(Jyotissastra)*
Astronomy is the science of heavenly bodies such as the Sun, the Moon, planets, stars etc. It includes   a) computing the longitudes of planets, stars etc.,   b) determining the time of eclipses,   c) measure- ments of various celestial bodies; and   d) their internal and external peculiarities etc.

AXIS (*aksha*)

Axis is an imaginary point around which the Earth spins (rotates) itself.

AYANAMSA

There are two systems of Zodiacs: 1) *Sayana*, with the first point of Aries (vernal equinoctial point), as the starting point and 2) *Nirayana* with a fixed initial point (star) in the ecliptic for measuring the celestial longitudes of planets etc. *Ayanamsa* is the distance between these two initial points. (There are different ayanamsas, as there is difference of opinion in the case of the year of coincidence of both the zodiacs.)

AYANAMSA, MEAN

Mean values of ayanamsa + precession in longitude for that date.

AYANAMSA, TRUE

Mean Ayanamsa + nutation in longitude.

CELESTIAL SPHERE *(Khagola, nabhomandala)*.

1. The celestial sphere is an imaginary sphere in the space, surrounding the Earth.

2. Earth is spherical and the surface of the Earth, if projected infinitely in the heavens, the figure formed will be celestial sphere.

COMBUSTION (astrlogy - *Moudya*)

The planets remain invisible to the naked eye for some days at the time of conjunction with the Sun. They become combust when coming very near to the Sun, that is, when they are within a particular distance from the Sun, eitherside.

COMETS *(Dhooma ketu)*

A comet which consists of a 'coma' (a shining nucleus) and unseen hydrogen clouds, is a star-like heavenly body, with a tail of light. It moves across the sky through regular routes. The tail which appears only when coming near the Sun, always keeps to the opposite direction to the Sun.

CONJUNCTION

When the apparent celestial longitude or Right Ascension of two bodies are the same, the phenomenon is called Conjunction.

CONSTELLATION

A group of stars, bearing the names of mythical Greek heroes and of objects occuring in Greek legends, to identify an area of the celestial sphere. (Lahiri's Ephemeris)

DAKSHINAYANA

It is the period from summer solstice to winter solstice, ie., southward passage of the Sun. Sayana solar months: Karkata to Dhanu. June 21 to December, 20. Consists of Rain, Autumn and Hemanta seasons.

DECLINATION

1. The 'declination' is the angular distance of a heavenly body from the celestial equator. It is positive or negative according as the celestial object is situated within the northern or southern hemisphere.

2. The declination of a heavenly body is its angular distance from the equator measured on an arc perpendiculr to the celestial equator drawn through the body.

DECLINATION OF THE SUN

At Vernal Equinox (March 21)   - 0°

Autumnal Equinox (September, 21) - 0°

Summer Solistice (June 21)   - 23° - 28' N

Winter Soistice (December 21) - 23° - 28' S

(The longitude is *sayana* and the declination is apparent, due to the movement of the Earth around the Sun.)

DIRECT MOTION

1. For orbital motion in the solar system, motion that is counter-clockwise in the orbit as seen from the north pole of the ecliptic.

2. Direct motion, for an object observed on the celestial sphere, is the motion from west to east, resulting from the relative motion of the object and the Earth (ie., forward motion of any celestial object).

ECLIPSES

1. Obscuration of a celestial body, caused by its passage through the shadow cast by another body.

2. An eclipse is the total or partial cutting off of the light of the Sun.

ECLIPSES, LUNAR

When the Earth comes in between the Moon and the Sun, ie., when the Moon passes through the shadow cast by the Earth, it results in the eclipse of the Moon. The eclipse of the Moon occur on full moon day (*Pournami*).

ECLIPSES, SOLAR

When the Moon comes in between the Earth and the Sun, ie. when the Earth passes through the shadow cast by the Moon, it results in the eclipse of the Sun. The eclipses of the Sun occur on new moon day (*Amavasya*).

ECLIPTIC *(Soorya marga)*

1. The ecliptic is the apparent path of the Sun in the sky. It passes exactly through the centre of the Zodiac longitudinally.

2. The mean plane of the Earth's orbit around the Sun or simply the apparent path of the Sun in the sky.

3. Ecliptic is the apparent annual path of the Sun amongst the fixed stars on the cosmic sphere. It is inclined at 23° - 28' to the celestial equator.

4. Ecliptic is a great circle on the celestial sphere whose plane passes through the Earth which is at its centre and it is the apparent yearly path of the Sun round the Earth.

EQUATOR, TERRESTRIAL *(Bhoo-madhya rekha, niraksha rekha, vishuvam)*

Eauator is an imaginary line round the Earth at equal distances from the north and south poles. It divides the Earth into northern and southern hemispheres. Geographically, equator is marked as 0° latitude. The north pole is 90° north and the south pole is 90° south from the equator .

EQUATOR, CELESTIAL

Celestial equator is a great circle of the celestial sphere, marked out by the extension of the Earth's equator.

EQUINOXES *(Ayanam, vishuvam, sama-ratri-dina)*

1. The two points on celestial sphere at which the ecliptic intersects the celestial equator. When the apparent longiude of the Sun is 0°, it is Vernal Equinox and when 180°, it is Autumnal Equinox. Day and night are equal on these points.

2. Twice a year the Sun crosses the equator. These are known as Vernal and Autumnal equinoxes (about March 20 and September 22 respectively). When the Sun reaches these two points, the day and night are equal in length. (Autumnal equinox = *Thula vishu*. Vernal equinox = *Mahavishuvam)*

FIRST POINT OF ARIES *(Meshadi)*

Since the zodiac is a circle with no beginning or end, an arbitrary point has been chosen to measure the 'signs' and also the longitudes of the planets. This point is known as the 'First Point of Aries' or Aries zero degree. As this is the point where the Sun enters northern hemisphere, it is also known as 'Vernal Equinox'.

GEOCENTRIC

1. Geocentric position of any heavenly body is the apparent position as seen from the centre of the Earh.
2. The observations with reference to the centre of the Earth is called Geocentric.

GRAHAS

The planets considered in Indian Astronomy for astrological purposes are called *Grahas*. The *grahas* are *Surya , Chandra, Kuja, Budha, Guru, Sukra, Sani, Rahu* and *Ketu.*

GREATEST ELONGATION

When the geocentric angular distances of Mercury and Venus are at a maximum from the Sun, it is called greatest elongation.

HELIACAL RISING AND SETTING

The planets Mercury to Saturn (as well as the Moon) remain invisible to the naked eye for some days at the time of conjunction with the Sun. The phenomenon of the planet's (and the Moon's) invisibility due to its proximity to the Sun is known as heliacal setting of the planets. The opposite phenomenon is defined as heliacal rising.

ECLIPSES

1. Obscuration of a celestial body, caused by its passage through the shadow cast by another body.

2. An eclipse is the total or partial cutting off of the light of the Sun.

ECLIPSES, LUNAR

When the Earth comes in between the Moon and the Sun, ie., when the Moon passes through the shadow cast by the Earth, it results in the eclipse of the Moon. The eclipse of the Moon occur on full moon day (*Pournami*).

ECLIPSES, SOLAR

When the Moon comes in between the Earth and the Sun, ie. when the Earth passes through the shadow cast by the Moon, it results in the eclipse of the Sun. The eclipses of the Sun occur on new moon day (*Amavasya*).

ECLIPTIC (*Soorya marga*)

1. The ecliptic is the apparent path of the Sun in the sky. It passes exactly through the centre of the Zodiac longitudinally.

2. The mean plane of the Earth's orbit around the Sun or simply the apparent path of the Sun in the sky.

3. Ecliptic is the apparent annual path of the Sun amongst the fixed stars on the cosmic sphere. It is inclined at 23° - 28' to the celestial equator.

4. Ecliptic is a great circle on the celestial sphere whose plane passes through the Earth which is at its centre and it is the apparent yearly path of the Sun round the Earth.

HELIOCENTRIC

The observations with reference to the centre of the Sun is called *Heliocentric*.

HORIZON (*chakravala*)

Horizon is the line at which earth and sky seem to meet.

LATITUDE, CELESTIAL

1. Celestial latitude (*kshepa*) is the angular distance of a heavenly body from the ecliptic.

2. Angular distance across the celestial sphere, measured north or south from the ecliptic, along the great circle, passing through the poles of the ecliptic and the object.

LATITUDE, TERRESTRIAL (*Akshamsa*: north--south)

Latitude of a place is its distance from the eqator, on its own meridian. It is reckoned in degrees from 0° to 90°, northwards or southwards. 'S' or 'N' is also given to show where the place lies.

LONGITUDE, CELESTIAL

Celestial longitude (*krantiamsa*) of a heavenly body is the angular distance of it measured along the Ecliptic, from the zero point (First point of Aries).

LONGITUDE, TERRESTRIAL (*Rekhamsa*: east--west)

The longitude of a place is its distance from 0° meridian (Greenwich) measured as an angle. It is expressed in degrees. ' E ' or 'W' is also indicated to show whether the place lies east or west of the meridian.

LONGITUDE, TROPICAL OR SAYANA

It is the distance between the first point of Aries and the heavenly body. It is measured along the ecliptic from Vernal Equinoctial point. As it is moving backward steadily and the longitude includes the *ayanamsa* also, it is called *Sayana*. It is the longitude as per the moving Zodiac.

LONGITUDE, SIDEREAL OR NIRAYANA

In Indian astronomy, the initial point from which the lontitude in the Zodiac is measured is a fixed point on the ecliptic. It is as per the fixed Zodiac and hence called Sidereal (based on stars) or *Nirayana* (as it is without *ayanamsa*).

MAANDI

Mandi or *Gulika* is an astrological *Upagraha*. It rises twice a day, that is, once in the day-time and once in the night, at fixed times.

M.C. (MEDIUM COELI)

The M.C. (Medium Coeli) is the point of intersection of the ecliptic with the meridian of the place. The longitude of this point is measured along the ecliptic from the 'Firtst Point ofAries'. It is the tenth house in astrology.

METEORS (*Ulkka*)

A meteor is a small body, rushing from the outer space to the atmosphere of the Earth, and becoming bright.

MERIDIAN (*Uchcha rekha*)

Meridian is an imaginary circle round the Earth, passing through a given place and the north and south poles.

MERIDIAN, CELESTIAL

Celestial meridian is a great circle passing through the celestial poles and zenit of a place. It is also called obesrver's meridian or Prime meridian.

MERIDIAN PASSAGE

Meridian is defined to be a great circle passing through the zenith and the celestial poles. It corresponds to geographical longitude of a place. A star or planet passes over the Meridian when the sidereal time of the moment becomes equal to the Right Ascension of the star or planet.

MERIDIAN CIRCLE

A great circle passing through the celestial poles and through the zenith of any place on earth.

NADIR

Nadir is the point of intersection of the clestial sphere with the plumb line produced downwards, ie., a point on the celestial sphere which is just below the observer's foot.

NAKSHATRAS (Lunar Mansions / Constellations)

The *nakshatra* is a system of diviion of the zodiac. In this system, the zodiac is divided into 27 equal parts and given the names of the constellations (group of stars) which are prominent within each division.

NODES (*Sandhi*, joint)

The nodes are the two points on the celestial sphere at which the plane of an orbit of any rotating celestial object intersects a reference plane. In respect of the Moon, these two points are known as Ascending Node and Descending Node (*Rahu* and *Ketu).*

NUTATION

Nutation is that part of the precessional motion of the pole of the Earth's equator which depends on the periodic motion of the Sun and the Moon in their orbits around the Earth.

OBLIQUE ASCENSION

The rising period of a sign is called 'Oblique Ascension' (*Rasimana* or *Rasi pramana).* It varies from place to place. The time of oblique ascension is computed for the signs of the tropical zodiac and *ayanamsa* is substracted from it to get *nirayana rasimana.*

OBLIQUITY

It is the angle between the equatorial and the orbital planes of a body. In Earth the obliquity of the ecliptic is the angle between the plane of the equator and the ecliptic.

OCCULTATION

Occultation is the obscuration of one celestial body by another of greater apparent diameter, especially the passage of the Moon in front of a star or planet.

OPPOSITION

When the apparent celestial longitude of two bodies differs by 180 degrees, the phenomenon is called Opposition

ORBIT (*Graha bhramana patham)*

Orbit is the regular path followed by an heavenly body round another. (Earth's orbit round the Sun.)

PERIGREE

Perigree is the point at which a body in orbit around the Earth is at the least distance from the Earth.

PERIHELION

Perihelion is the point on a planetary orbit when it is at the least distance from the Sun.

PLANETS

1. The heavenly bodies which go round the Sun are called planets.

2. The planets are the heavenly bodies which rotate about themselves and revolve round the Sun along their own paths called orbit. The planets are Mercury, Venus, Earth, Mars, Jupiter, Saturn, Uranus, Neptune and Pluto.

3. A fixed star appears twinkling while the planets shine with steady light.

PLANETS, EXTRA-SATURNINE

The planets whose orbits lie outside the orbit of Saturn are called extr-Saturnine planets. (Uranus, Neptune and Pluto).

PLANETS, INFERIOR OR INNER

The planets whose orbits lie between the Earth and the Sun are called inner, interior or inferior planets. (Mercury and Venus)

PLANETS, MINOR

These are the fragments of a broken planet, between Mars and Jupiter. Also called Asteroids.

PLANETS, OUTER OR SUPERIOR

The planets whose orbits lie outside the orbit of the Earth are called outer, exterior or superior planets (Mars, Jupiter, Saturn etc.).

POLES (dhruva)

The poles are the two ends of the Earth's axis., ie., north pole and south pole.

POLES, CELESTIAL

If the axis of the earth is extended infinitely it will cut the celestial sphere at two points which are known as celestial poles.

PRECESSION OF THE EQUINOXES (Ayana chalanam)

At the time when the Sun reaches the equinoctial point at Aries 0^0 , the position of the Earth, with reference to fixed stars, is nearly $50\frac{1}{3}$ " west than that of previous year. This westward motion of the zodiac by $50\frac{1}{3}$ " a year, causes a slight movement in the case of the equinoxes also. This motion of the equinox is known as *precession of the equinoxes.*

PRECESSIONAL DISTANCE (Ayanamsha)

It is the distance (in degrees, minutes and seconds) between the First Point (starting point) of the two zodiacs, namely, the Sidereal (or fixed) and Tropical (or moving) Zodiacs.

RAHU AND KETU

Rahu and Ketu are two nodal points where the path of the Moon cuts the path of the Earth. When the Moon crosses the ecliptic, in the course of going from south to north, it is Rahu (Ascending

Node).   Similarly when the Moon crosses the ecliptic, going from north to south, it is Ketu (the Ascending Node).

RETROGRADE MOTION

It is the motion which is clockwise in the orbit, as seen from the north pole of the ecliptic. For an object observed on the celestial sphere, it is the motion from east to west, resulting from the relative motion of the object and the Earth.

RERTROGRESSION (astrology - *Vakra*) :

Retrogresion is the backward motion of the planets. (Actually the planets never go backward but due to the movements of the Earth and the other planets, around the Sun at different rates of speed, occupying different angles, it so appears.)

RIGHT ASCENSION

1. 'Right Ascension' is the angular distance on the celestial sphere measured eastward along the celestial equator from the *Vernal Equinox*, to the *hour-circle* passing through the celestial object.

2. The right ascension is the angular distance between the first point of Aries and an arc perpendicular to the celestial equator drawn through the body.   The first point here is Aries 0^0 or Vernal Equinox. In other words, *sayana* longitude is considered for this purpose.

SATELLITES (*Upagraha*)

1. Satellites are heavenly bodies moving round the planets. The Moon is the satellite of the Earth. The Mars has 2 satellites, Jupiter 16, Saturn 17, Uranus 5, Neptune 4.   There can be more sattellites yet to be discovered.

2. Satellites are those heavenly bodies which move around the planets and in turn move around the Sun alongwith the planets and are normally called moons of the planets, like the Moon which is a satellite of the Earth.

SIGNS OF THE ZODIAC

The zodiac is divided into twelve equal parts called 'Signs' (30^0 x 12 = 360^0·).   These twelve divisions are called Aries, Taurus etc.   In the case of the tropical zodiac, each of these signs formerly occupied the similarly named constellations,   but now, by precession of the equinoxes, they coincide with the constellations that bears the names of the preceding signs, that is, the constellation Pisces is now in Aries Sign.

SOLAR SYSTEM (*Soura yoodha*)

Some heavenly bodies like the Earth go round the Sun. These are planets, their satellites, asteroids and comets.   All these bodies with the Sun is called Solar System.

SOLSTICES

The times of solstices are those at which the Sun's apparent longitude is:
a) 90 degrees, Summer Solstice (end of *Utharayana*); and
b) 270 degrees, Winter Solstice (end of *Dakshinayana*)
At summer solstice, the day is longest and at winter solstice, the day is shortest for observers in the northern hemisphere.

SQUARE

When the apparent celestial longitude of two bodies differs by 90 degrees or 270 degrees, the phenomenon is called Sqare.

SYSTEMS, SAYANA AND NIRAYANA

The system of astronomy/astrology which follows the moving zodiac is called " Tropical (*Sayana)"* and the system which follows the fixed zodiac is "Sidereal *(Nirayana)"*. Western astrologers and modern astronomy follows *Sayana* system. Indian astrology follows *Nirayana*.

TRANSIT

The transit time of Sun, Moon and planet is the time of its entrance into the signs or *nakshatra* divisions of the Zodiac.

UNIVERSE

The sky which is visible from the Earth is actually that part of the Universe which enfolds this planet. The groups of stars which appear like luminous islands in the sky are Galaxies. There are more than ten thousand crores of galaxies which contain crores of stars. The Milky Way Galaxy *(Ksheera patham, Akasa Ganga)* is one among them. The Sun is a Star in that galaxy.

UPAGRAHAS

The *Upagrahas* are secondary planets in Indian astrology. They are not physical bodies but mathematically computed astronomical points on the ecliptic with reference to the Sun's longitude. The important *Upagrahas* are Dhuma, Patha, Paridhi, Yamakantaka etc

UTTARAYANA

The northward passage of the Sun from winter solstice to summer solstice. The sayana solar months from Makara to Mithuna. December, 21 to June, 20. Consists of Winter, Spring and Summer seasons.

VERTICALS

Great circles drawn perpendicular to horizon from the zenith are called verticals. These are also called secondaries to the horizon.

YEAR OF COINCIDENCE

It is the year when the starting point of the two zodiacs, that is, the moving and fixed zodiacs was one and the same.

ZENITH

Zenith is the point of intersection of the celestial sphere with the plumb line produced upwards, ie., a point on the celestial sphere which is vertically above the observer's foot.

ZODIAC *(rasi chakra, rasi mandala)*

1. Zodiac is an imaginary belt of about 9^0 North and 9^0 South of the ecliptic where the moon and all other planets have their movement.

2. The Zodiac is an imaginary belt in the heavens, stretching about 8° - 9° either side, ie., north and south, of the ecliptic, within which the planets move around the Earth, once in a day, from east to west. This east-west movement of the Zodiac is apparent. It is due to the west-east rotation of the Earth on its axis.

ZODIAC, SIDEREAL

A fixed zodiac is considered in Indian astronomy for astrological purposes. In this case, the first point or 0° is a fixed star. As it is based on stars, it is called 'sidereal' or 'fixed' zodiac.

ZODIAC, TROPICAL

The zodiac which reckons the Aries 0° from the equinoctial point and which has precession, is the 'movable' or 'tropical' zodiac.

mullappillymp@gmail.com
28th March 2013

r.

www.ingramcontent.com/pod-product-compliance
Lightning Source LLC
Chambersburg PA
CBHW080607180526
45168CB00007B/2810